Teaching Biblically-Based Science

How to Establish Your Science Classes on the Foundation of God's Word

A Workbook

Christa Jewett

ISBN-13: 978-0615817606

As the rain and the snow come down from heaven, and do not return to it without watering the earth and making it bud and flourish, so that it yields seed for the sower and bread for the eater, so is my word that goes out from my mouth: It will not return to me empty, but will accomplish what I desire and achieve the purpose for which I sent it.

– Isaiah 55:10-11 NIV

Table of Contents

Part 1: Establishing a Biblical Foundation for Your Science Lessons

Part 2: Addressing the Creation Versus Evolution Issue

Part 3: Practical Tips for Your Science Class

Part 4: In Summary

Foreword

When the Lord led me to begin teaching biblically-based science education through Saltwater Studies in 2011, I found the prospect exciting but overwhelming. Because Saltwater Studies was a new work of the Lord, I felt led to write my own curriculum. It was my desire to prepare curriculum that is based on the Word of God and offer classes that would glorify Him. Initially, I found the prospect of preparing my own curriculum to be daunting. As a result, I have spent many hours seeking Him for direction and understanding as to what He would like for me to teach and how He would like me to present scientific information.

During this time, the Lord has been abundantly faithful, revealing to me so many aspects of Himself through His Creation, which I, in turn, am excited to share with my students. Not only are my classes a learning opportunity for my students but they are also a learning process for me as the Lord continues to give me greater insight into His character and personal nature. This process has deepened my relationship with Him and has also helped me to use my science classes to encourage my students to enjoy a personal intimacy with their Lord and Savior Jesus Christ through His Creation. I have been so blessed by this process and excited at what the Lord has taught me that I have started to share what I have learned with homeschooling moms. Consequently, the Lord has led me to prepare this book to offer homeschooling moms the opportunity to share in the information that the Lord has revealed to me.

While this book is presented to homeschooling moms, it is meant to be useful to anyone who is interested in applying these principles in their own lives or in their interactions with children. This book is written in a workbook format with questions included at the end of each section to offer a time of reflection as you consider the principles discussed. My prayer is that this workbook will initiate for you a new phase in your journey of faith that will continue as you grow in your knowledge of Jesus Christ. Since our God is infinite, there is no end to the number of things that we can learn about Him! As your relationship with Him is deepened it will place you in a better position to encourage your children on their own journeys of faith. I am excited to continue this journey with you as we look to our Lord for greater understanding of His character through what He has made.

Christa Jewett

PART I

Establishing a Biblical Foundation for Your Science Lessons

What is Biblically-Based Science?

When many people hear the phrase "biblically-based science" they think that it means teaching their class from a Creation-based viewpoint instead of teaching evolutionary theory. However, biblically-based science is so much more than substituting "God created" for "We evolved". While Creation is a critical part of a biblically-based science class, the Bible also includes a wealth of information about our world, its origins, how it operates under the care of our Lord and our responsibilities as caretakers of the earth. For example, a brief review of a few passages of Scripture reveals that:

- The triune God (God the Father, God the Son and God the Holy Spirit) created the world (Genesis 1:1; Colossians 1:16).
- The Lord established mankind as caretakers of the earth (Genesis 2:15).
- The earth belongs to the Lord (Psalm 24:1-2).
- All of Creation testifies of its Creator (Romans 1:20).
- We can learn about the character of our Lord by investigating what He created (Psalm 19:1-2).
- Observing Creation gives us an opportunity to worship the Lord (Romans 1:20-25).
- The Lord sustains the world (Psalm 3:5; Colossians 1:17).

Biblically-based science is much more than substituting "God created" for "We evolved".

A well-balanced biblically-based science class should include these truths at minimum as the foundation for learning and understanding the world around us. Consequently, we and our students will obtain a more complete picture of God's involvement not only in the world but also in our lives. If we only focus on defending the Christian faith against the theory of evolution we are missing a large portion of what science can reveal about the Lord. And as any sports enthusiast knows, the key to winning a game is having both a strong defense and a strong offense. If we develop our science classes so that we only focus on the defense of Creation, where is our offense where we teach our students the remainder of Biblical truth? It is time that we, as followers of Jesus Christ, take the initiative to teach our children *all* of what the Bible has to say about Creation and our natural environment.

Rather than recreating the Creation versus evolution debate, this workbook will focus more on the principles behind biblically-based teaching and the Creation versus evolution issue. I recommend that you seek out other resources that provide a complete scientific defense for Creationism in order to ensure that you have a well-balanced approach in your science class. While there are numerous wonderful ministries that tackle this critical issue, two that I highly recommend are: Answers in Genesis (www.answersingenesis.org) and the Creation Studies Institute (www.creationstudies.org).

Another critical point as to why we must include the entire counsel of the Word of God in our science classes is to protect our children against future attacks of the Enemy. Research indicates that nearly three out of every five (59%) young Christians leave the church permanently or for a long period of time after the age of 15. In 2011, David Kinnaman, president of the Barna Group wrote a book summarizing the results of a five-year project that investigated why so many students are leaving the church. His results highlighted six major reasons given by these students for leaving the church. Two of these include that their experience of Christianity is shallow and that churches come across as antagonistic towards science. Twenty percent of these students made the alarming statement that "God seems missing from my experience at church."[1] A shallow relationship with God and scientific confusion are two facets that are playing a significant role in students falling away from the faith. God has placed you, as a homeschooling mom, in an ideal position to address both of these issues for your children. Rather than defending the Christian faith against their questions in the future, you can use your science classes to help them interact with God on a personal level and prevent Satan from using scientific questions to undermine their faith when they are older.

One of the great advantages of being a homeschooling mom is that you have ample opportunity to teach the Word of God to your children. Your time is not spent on combating lies and misinformation propagated at secular schools but can instead be focused on what is true. In addition, by weaving the foundational truths of the Bible through your lessons, your children will learn that biblical faith extends to all areas of their lives. Science in particular is an excellent subject where you can include the Lord and the Word of God in everything that you do. However, I know that many homeschooling moms are hesitant or unsure as to how to teach science to their children. I have learned that homeschooling moms have a wide variety of perspectives about their science classes. Perhaps you can relate to one of these statements:

- "Science is just not my thing. I do not really focus that much on it at home because my children think it is boring. I feel like I am pulling teeth whenever I am trying to make them do their science lessons."
- "I teach science using a written curriculum but I am always looking for ideas as to how I can make the classes fun and interesting for my children."

- “I find it difficult to teach science to my children because it intimidates me.”
- “I was raised believing that evolution is true. I know what the Bible says but I still do not really know what to teach my children about science.”
- “I feel inadequate when I teach science because I worry that I am not going to have the answers to all my children’s questions.”
- “I just do not understand science and it overwhelms me. I know that I need to include it in my children’s studies but I just do not know how to do it.”

If you can relate to any of these statements, I hope that this workbook will give you a fresh perspective on teaching science in your home. I pray that you will be encouraged, energized and ready to try a new approach with your children that will glorify the Lord.

Personal Reflection & Application

1. What do you want to accomplish by completing this workbook?

2. What is your current opinion or thoughts about teaching your science class?

3. What do you want to change about your science class?

4. What has the Lord revealed to you by reading this section?

Focus Your Science Classes on Jesus Christ

In order to ensure that your science classes are honoring to the Lord, you must establish them on the foundation of Jesus Christ. Just as what we believe about Jesus Christ impacts our lives for eternity, what we believe about Him should also influence every aspect of our science class. But what do Jesus Christ and science have to do with each other? What does the Bible say about Jesus Christ and His relationship with Creation? According to the Word of God, Creation was made by Jesus Christ, for Jesus Christ and through Jesus Christ. He is the source of all living things.

For by Him all things were created that are in heaven and that are on earth, visible and invisible, whether thrones or dominions or principalities or powers. All things were created through Him and for Him. And He is before all things, and in Him all things consist.

– Colossians 1:16-17

Since science is the study of the natural environment of which Jesus Christ is the literal source, our science studies and lessons must include Him. In addition, by saying that in Him all things consist, Colossians 1:17 is also saying that Jesus Christ literally holds us together. In Psalm 3:5, King David offers praise to the Lord, acknowledging that we wake every morning because He sustains us. The Lord holds our very lives in His hands.

Jesus Christ is central to Creation and our very existence so all of our scientific studies must include Him to provide an accurate representation of our natural environment.

Without Jesus Christ, we cannot even offer an answer to the most basic of scientific questions: "Why does life exist?" Those who believe in evolution, in particular, cannot answer this simple question. Evolutionary theory, at its core, removes the need for God, thereby eliminating the source of life. Consequently, evolutionists are prevented from being able to offer a viable answer to this question. One of my favorite worship songs, "You Are" describes a number of Jesus Christ's wonderful attributes. It includes a line that says, "You are every question's answer. You are every reason why." The truth in the Bible echoed in this song underscores that there is no way to offer an explanation for the existence of life in our world apart from Jesus Christ. Jesus Christ must be included in our science lessons, so that we can offer our children answers not just to what exists but also how and why all Creation exists. Without Jesus Christ, life just does not make sense.

Personal Reflection & Application

1. What has the Lord revealed to you by reading this section?

2. What role does Jesus Christ play in Creation?

3. Why is it important to include Jesus Christ in our science classes?

Understanding the Purpose of Creation

One of the most important Creation-related verses in the Bible is Romans 1:20. In this verse, we find that one of the reasons God created the world was to testify of Himself.

> For since the creation of the world His invisible attributes are clearly seen, being understood by the things that are made, even His eternal power and Godhead, so that they are without excuse.
>
> – Romans 1:20

This verse indicates that God's invisible attributes, including His eternal power and divinity, are all revealed through His Creation, and have been since the day the earth was created.

> *One of the main purposes of Creation is to reveal information about our Creator.*

According to the Word of God, Creation reveals everything *all* people need to know about Him, eliminating any potential excuse for them to live apart from Him. The next few verses of Romans 1 continue on to say that many of the sins present in our world today are a result of people seeing evidence of God in Creation but refusing to honor or worship Him. Because people have chosen to worship idols rather than the Lord, He has allowed them to be overtaken by their sinful desires which will cause them to face consequential judgment. Our science lessons have eternal consequences. By teaching our children to look for the character of God in Creation and to honor Him for what

He has made, we are helping them to have a relationship with Him that can save them from eternal judgment.

So how we can obtain information about our God from Creation? Like a master painter signing a piece of artwork, the Lord has left His signature on everything that He has made. As a potter shaping a vase leaves his fingerprints in the clay he uses, so the Lord has left evidence of Himself throughout our entire world. These "fingerprints" appear in a number of different ways. One of these is that we see similarities throughout plants and animals in both form and function. For example, bumble bees and birds both fly using wings, a similarity in function, although their wings look dramatically different. Does this mean that they might be somehow related in an evolutionary tree? No, quite the opposite. Instead, these features point to a common Creator who enjoys making animals that fly. As another example, there are some species of crabs, shrimp and flies that have eyes with similarities in form. A number of these species have stalked eyes with a 360 degree field of vision that can move independently of each other. Does this mean that crabs, shrimp and flies are all somehow related to each other? No, rather, this is evidence of a common Creator who designed an amazing feature and reused it in different species within His Creation. Just as an engineer might repeatedly use well-developed technology in different products he is working on, the Lord uses similar beautiful and well-developed designs in different species He has created.

Another "fingerprint" of God present in Creation is that we see evidence of His character in what He created. Like a child that has inherited genes from his parents, both plants and animals display the same character traits that we see in our God. Later in this workbook, I will discuss examples as to how we can see the attributes of God displayed in various animals and plants. We need to use our science classes to teach our children and students both the facts of science and how to recognize the corresponding characteristic of God. By training our children and students to see evidence of God in all that He has made, we will fill their lives with truth that will enable them to reject the lies of the Enemy later on in life.

Personal Reflection & Application

1. What has the Lord revealed to you by reading this section?

2. What is one of the main purposes of Creation?

3. How can we see the character of God through His Creation?

A Display of God's Personal Love

So, why would the literal Master of the Universe go to so much trouble to leave evidence of Himself in everything that He made? The answer is simple. Creation reveals God's amazing love for us and His desire to have a personal relationship with each of us. My personal belief is that this world is the first of two love letters that God has given us with the second one, of course, being the Word of God. Our God, who spoke the entire Universe into being, created all that we see so that we can have a relationship with Him.

God's love for us, as revealed through Creation, is deeply personal and intimate. In Psalm 139, King David describes the Lord's vast and intricate knowledge of every one of His children. He writes that the Lord formed us in our Mother's womb and knows the number and substance of our days before we were even born. But rather than describing a cold and distant relationship that might be expected between an Engineer and His created work, this chapter describes the love and fondness between a Father and His children.

"How precious also are Your thoughts to me, O God! How great is the sum of them! If I should count them, they would be more in number than the sand."

– Psalm 139:17-18a

According to these verses, our Father thinks of us more often than the grains of sand in all the beaches of the world. Even more importantly, His great and amazing love is displayed freely through His Creation!

One of the ways that I have seen the Lord display His love is by interacting with His people on a personal level through His Creation. Since the Lord knows each of us intimately, He knows exactly what will bring us the greatest enjoyment in His Creation. I have seen the Lord frequently modify the wildlife seen during my field trips to respond to the likes and interests of those attending. Although I may return to the same beach numerous times, each time, I know that we are going to see something different as the Lord takes the opportunity to interact personally with my students and their parents. He loves to bless His children in this way.

One field trip in particular stands out in my mind because of its unique circumstances. I had scheduled a field trip to Marco Island on the west coast of Florida to collect sea shells and had several students signed up to attend. Unfortunately, a hurricane made landfall on the Gulf Coast the day before the scheduled field trip and the winds were still too strong to be out on the beach. I knew that my students would be disappointed so as a last minute alternative, I suggested a beach close to Fort Lauderdale on the East Coast of Florida where the winds would be less severe. Two families chose to attend. As we made our way slowly down the beach, we started to find numerous shells washed ashore by the storm. I found myself talking to one of the women who attended and learned that her family had come to the beach with me to get out of their house because their roof had developed a major leak during the storm. The leak flooded one of the main rooms, soaking their drywall and ruining numerous household and personal items. In order to take a brief break from the mayhem and clean-up, they decided to come to the beach with me. While this situation represented a difficult trial for their family, they were trusting in the Lord that He would provide for their needs. We continued along the beach and the conversation shifted to the sea shells we were collecting. I started to discuss different features of the mollusks (snails and clams) we were finding on the beach.

At a pause in the discussion, the woman with whom I had been talking commented to me, “I just love the snails with holes in the sides so that you can see the inside of the shell. I think the insides of the shell are so beautiful.”

As with everything in life, I have learned that people have different preferences in regards to sea shells. Personally, if I found a sea shell with a hole in it, I probably would not keep it because I generally collect shells for the purpose of teaching my students to identify them. The more flawed the shell, the more difficult it is for a student to identify. Consequently, I only save shells that are as close to perfect specimens as possible. Shortly after the woman’s comment, I paused to help a student identify a shell she had found on the beach. The woman continued on, looking for more sea shells. Only a few minutes later, she ran back down the beach toward me with a huge smile on her face.

The Lord loves His children. He knows us intimately and loves to give us gifts, knowing exactly what will be the greatest blessing to each of us.

“Look what I found!” She shouted. “Isn’t it just beautiful?”

In her hand she was holding a large Apple Murex shell, a beautiful type of snail. Murex snails are the supermodels of the mollusk world. God created them with a flair for the dramatic as they often display elaborate ridges, spines, and folds on their shells. The shell features are combined with beautiful white to sunset-like coloring that make them highly desirable for collectors. This Apple

Murex was sporting caramel, tan and brown stripes as well as a deep purplish pink interior. It was about three inches long, the largest one I had ever seen found on the beach by an amateur collector. In its side were two perfectly placed holes that nicely displayed the interior of the shell.

"And look, it has holes so that I can see the inside!" She exclaimed. "I love it!"

The Lord knew that of all the people on our trip, this woman would be the one to enjoy this particular shell. If I had found it, I would have been disappointed that such a beautiful specimen was "ruined" by the holes in the side. In His infinite wisdom and love, He allowed this woman to find this shell, blessing her in the midst of her trial by giving her just the right gift that would bring her excitement and pleasure. We serve a personal God who loves His children.

Scenarios such as these occur frequently on my field trips. I now ask the Lord verbally at the beginning of my field trips to give each of my students a personalized gift so that they will recognize His blessings designed specifically for them. This is such a wonderful opportunity for children to interact with the Lord Jesus on a personal level through His Creation. This lesson is so much more important than memorizing facts or learning the scientific method. God is not a cold and distant Engineer but an ever-present Father who greatly desires an intimate relationship with His children. I offer you a challenge: the next time you go on a field trip with your children, pray with them at the start of the field trip and ask the Lord to show you something special during your time in His Creation. When you seek to honor God in your science classes, prepare yourself to enjoy displays of His abundant love for you. He has personalized blessings waiting for you and your children!

Personal Reflection & Application

1. What has the Lord revealed to you by reading this section?

2. How can we see God's love for us in Creation?

3. Why does God so clearly display His love for us in what He has made?

A World We Can Understand

Another way that we can see God's love for us is that He created a world we can understand. Our God is infinite in wisdom and in knowledge. Yet instead of creating a world that is consistent with His level of understanding, He created one that even we, finite beings, can comprehend. Regardless of any single person's intelligence level, God in His love for us, made sure that there is something in this world that he or she can grasp. In the display of this ability, we can glimpse a shadow of the magnificent capabilities of our God. As an engineer will tell you, it is often the simplest designs that are the most brilliantly constructed. Greater talent is often displayed in simplicity rather than complexity. In his biography of Giotto di Bondone, an artistic genius of the Italian Renaissance, Giorgio Vasari relays a story of how Pope Boniface VIII requested copies of Giotto's work. In response, Giotto dipped his brush in red paint and painted a perfect circle with one continuous stroke. This perfect circle was sufficient for the Pope to conclude that Giotto was the greatest painter of his time.[2] Just as the simple circle testified of the extraordinary talent of Giotto, so our world testifies of the immensity of our God.

> *God created a world that we could comprehend, well below His own knowledge and understanding, so that we can have a relationship with Him through His Creation.*

On the other hand, the opposite is also true. Just as God designed a world with the purpose of creating something that everyone, of every intelligence level and of every gifting, could understand, there are also going to be elements of our world that are beyond our comprehension. It does not matter if you are the most intelligent person on the planet, there are still going to be aspects of this world that you do not understand because no one is capable of completely understanding an infinite God with a finite mind. First Corinthians 3:19 says, "The wisdom of this world is foolishness with God." Although God has created some people with intelligence that is well beyond mine or yours, it is important for us to remember that our God is infinitely greater. Some secular scientists may try to dazzle us with their theorems and logic but ultimately, without the Lord, even their greatest thoughts and accomplishments are foolishness to Him.

Watch your children's interest in science grow as they learn that enjoying Creation is a way to interact with our Savior.

If you are one of those people who feels uncomfortable or intimidated by science, this truth should be a great encouragement to you. We must remember that God is the author of *all* knowledge and understanding. He also gives liberally to those who ask Him for wisdom (James 1:5). If there is anything that you are having difficulty understanding, ask Him to grant you wisdom. He will be faithful to answer you because He *wants* you to be able to understand His Creation for the purpose of deepening your relationship with Him. This truth is also a great way to encourage your children who struggle with science or have other gifts and interests. Their interest in science will grow as they learn that enjoying Creation is a way to interact with their Savior. And since our God is infinite in knowledge, we can expect to continue to grow in understanding of Him as we seek to include Him in our science classes.

Personal Reflection & Application

1. What has the Lord revealed to you by reading this section?

2. What does it mean to you that God created a world we could understand? How does this influence your perspective of Him?

Displaying Active Faith

As both your child's teacher and parent, you are responsible for setting an active example of a thriving faith in Jesus Christ. That includes proclaiming the name of Jesus Christ throughout your science lessons. I learned this valuable lesson on one extremely hot summer day in South Florida during a field trip at a local park. The humidity made the air feel heavy; draping suffocating heat over everything it touched like a hot heavy blanket. It was the kind of humidity that caused you to start sweating the second you stepped out of an air conditioned building. And I had a field trip scheduled - a guided hike through a mangrove forest. Since I had a good size group attending, I did not want to cancel as rescheduling always proves to be difficult due to busy schedules and other commitments. I filled spray bottles with cool water, hoping that this might offer my students some relief during the walk. By the time the field trip began at 9:00 AM, I knew most animals would be hiding in the shadows to avoid the oppressive heat. However, throughout the course of leading field trips, the Lord has taught me that He is in control of all that we see. He had always shown Himself abundantly faithful on all of my prior field trips. We always see such amazing things because the Lord blesses those who go out in His name. So regardless of the weather, I continued with the field trip, trusting in Him that He would bless us as He always had on prior trips.

As both your child's teacher and parent, you are responsible for setting an active example of a thriving faith in Jesus Christ.

As we entered the mangrove forest, the little breeze that had offered a slight relief from the heat was quenched by the throng of mangrove trees growing in tangled bunches. I led the group along the boardwalk pointing out different features of the mangrove forest, my eyes searching the trees and water for any signs of life. A bird, a crab, a fish, anything! I caught sight of a few mangrove crabs hiding in some tree roots and quickly pointed them out. I hurried the students along to one area of the mangrove forest that I call the Puffer Hole in hopes of showing them something a little more interesting than a small mangrove crab. The Puffer Hole is a relatively open area adjacent to the walkway where checkered puffer fish live in the roots of several trees. These fish were consistently at this location every prior field trip I had led at this particular park.

I said a personal prayer as we walked, "Lord, You know that we are here in Your name. We are here to give You praise for Your beautiful Creation. Please show us something special today that we might give You glory. Please show these students how You bless those who come in the name of the Lord. Please help those puffer fish to be at the Puffer Hole."

When we reached the Puffer Hole, I had all the students line up along the railing and look for the checkered puffer fish that I had prayed about. We waited but no fish appeared. The Lord had always been so faithful to me by revealing special aspects of His Creation to my students. I did not understand why He was not offering us a blessing on this day. Finally, without seeing any of the previously reliable checkered puffer fish, we continued with the hike. At this point, my students were extremely hot and starting to become whiny. I tried to keep the field trip upbeat and fun without revealing my disappointment. I just did not understand why the Lord was not showing us more wildlife on this day. I took my students all the way to the end of the boardwalk to no avail. We did not see any more wildlife.

"Well, for some reason, the Lord had a different plan for us today," I thought to myself.

As we walked back to the start of the hike, retracing our steps along the boardwalk, I almost passed the Puffer Hole once again without stopping. Not wanting to give up on potentially seeing some sort of animal, I stopped the students to have a second look.

I again breathed a prayer to the Lord, "Please, Lord, send out these puffer fish."

No puffer fish. For some reason, I felt led to repeat my prayer out loud.

I said again, loud enough for all of my students to hear, "Lord, we would really like to see the checkered puffer fish. Can you please bless us and send him out?"

Before I could finish speaking, a checkered puffer fish swam out of the roots directly in front of my students. Leaving the shadow of the trees, it swam in a straight line towards them, stopped right beneath where they were standing, then turned around and headed straight back into the roots. His swim was reminiscent to me of a model on a catwalk making a special appearance. My students shouted excitedly and I cheered, knowing that the Lord had sent Him out specifically for us in answer to my prayer. The Lord, in His great mercy, had rescued my field trip.

The valuable lesson that I learned that day is the need for us as teachers to be active role models in living lives of faith in front of our children and students. My students did not know the prayer that I was saying in my heart as we walked along the boardwalk. And if the Lord had answered my personal prayer, they would have missed out on learning a valuable lesson about the Lord and His faithfulness. Through this situation, the Lord started to teach me the importance of proclaiming His name to my students and involving them in the process of faith so that they can learn more about Him.

Since this field trip, I have had experiences where even after offering a verbal prayer to the Lord for His blessing, He still did not answer my prayer right away. Rather than this being the Lord saying "no" to my prayer, I have learned that He is saying "not yet" because He is waiting for an opportunity to interact with my students instead. I had a recent situation where I was on a field trip to Marco Island with one of my younger students. As we walked along the beach looking for sea shells, my student talked excitedly about finding a sand dollar. However, it was high tide and the prospect of finding sand dollars in breaking waves is rather low apart from the Lord. After walking nearly two miles down the beach, we found a relatively calm, shallow area on a sandbar and I offered a verbal prayer to the Lord asking Him to help us to find a sand dollar. Normally, the Lord responds immediately to my requests so I was surprised when we did not find a sand dollar right away. My thoughts turned inward toward my motivations.

"Am I trying to manipulate God into doing what I want Him to do by following a 'formula' for prayer that has worked in the past? Or am I asking with a genuine desire for Him to be glorified?" I asked myself.

As I pondered these thoughts, my eyes continued to scan the water. A few minutes later, I gratefully spotted a sand dollar partially buried in the sand. Excited, I called my student over to pick it up. As he grabbed it from the sea floor, he grinned from ear to ear.

"See how the Lord blessed us by answering our prayers?" I was quick to point out to him.

"I know," my student responded. "I just asked Him to help me find one."

In this moment, the Lord showed me that He did not immediately answer my prayer because He wanted the opportunity to respond to my student's prayers. If God appears to be silent, be patient. He is faithful.

He may just be waiting for your children to offer a prayer of faith so that He has a chance to respond to them directly.

Ultimately, our science classes are both about our faith and the faith that we are seeking to instill in our children and students. Without these valuable lessons, science merely becomes head knowledge without any eternal significance or impact in our children's lives. We need to use every opportunity we have to prepare our children for today, tomorrow and eternity.

Personal Reflection & Application

1. What has the Lord revealed to you by reading this section?

2. How can you display a life of active faith in your science classes?

3. How can you involve your children in a display of active faith during your lessons?

Science Class is an Opportunity for Worship

As you seek to establish a biblical focus for your science classes, a critical aspect of your lessons must be worship. This is because another purpose of Creation is to give glory to the Lord. We as a part of God's Creation were created to praise Him and bring Him glory.

Let everything that has breath praise the Lord. Praise the Lord!

– Psalm 150:6

I often share a joke with my students and tell them that instead of saying that they are attending a science class, they should tell their friends that they are coming to a worship service. This is because science class done correctly becomes a worship service. Unfortunately, our church culture today may lead Christians to believe that worship is limited to a few songs on a Sunday morning. However, worship is a lifestyle that must be cultivated in our children. As we seek to know the Lord through His Creation and He is faithful to reveal Himself to us, we have the responsibility to respond with thanksgiving and praise. During your classes and field trips this means appreciating the Lord for what He shows you verbally in front of your children and students.

Worship is a lifestyle that must be cultivated in our children.

Comments like, "Oh wow! That is beautiful! Thank you, Lord for your beautiful Creation!" need to be repeated over and over in order to help cultivate a lifestyle of worship in your children and students. They need to learn to direct their praise to the Creator rather than the Creation. Unfortunately what we see today in our society is that people often misdirect their worship, placing the priority of the Creation over the Creator. It is our responsibility to make sure that our children and students direct their worship appropriately to their Lord and Savior Jesus Christ.

In addition to sprinkling praise-filled comments throughout your science classes, I also encourage you to engage in a time of praise and thanksgiving with your children at the end of your field trips or science classes. This is something that I started doing in my own field trips in response to the Lord's conviction. Prior to initiating this time of worship, I would pray collectively with the students at the beginning of the field trip and ask for the Lord's favor and blessing. But then when the Lord was faithful to answer my prayer, I would not collectively thank Him or praise Him again with the students at the conclusion of the field trip. One day as I was thinking about my field trips, the Lord reminded of the account of the ten lepers that Jesus healed in Luke 17:11-19. Although all ten lepers were healed, only one returned to say thank you and worship Him. The Lord showed me that I was acting similarly to these ungrateful lepers who begged for mercy and then went on their way without praising the One who healed them. In response to the Lord's conviction, I made a commitment to conclude future field trips with a collective time of praise to the Lord for His faithfulness.

While a collective time of worship and praise at the conclusion of a field trip might seem rather simple, for me it represented a giant step out of my comfort zone. My personal worship preference is rather conservative so I was nervous about implementing this with my students. With the date of my next field trip nearing, I waffled on my commitment to the Lord. As I thought about my fears, the Lord reminded me of King David worshipping before Him as the Israelites brought the Ark of the Covenant into the tabernacle.

"Then David danced before the Lord with all his might; and David was wearing a linen ephod. So David and all the house of Israel brought up the ark of the Lord with shouting and with the sound of the trumpet."

– 2 Samuel 6:14-15

David's behavior even brought ridicule from his wife, but his reply was simple and clearly displayed his heart for the Lord. In response to Michal's criticism, he said, "And I will be even more undignified than this, and will be humble in my own sight" (2 Samuel 6:22). The Lord appreciates and honors the heart of a humble worshipper. My concerns were nothing more than the reflections of a prideful heart.

In my next field trip, the Lord once again was abundantly faithful to my students and me, showering us with blessings. We enjoyed beautiful weather on the beach and an amazing display of sea life that instilled a sense of excitement and wonder in my students. In accordance with my commitment to the Lord, at the conclusion of the field trip, I gathered my students for a collective time of praise and thanksgiving to the Lord. The students and I gave Him a round of applause while I shouted "Praise the Lord for His blessings!"

After we were done clapping, one of my students blurted out, "Well I've never done *that* before."

Regardless of the fact that I felt awkward in trying something like this for the first time, my student's response indicated to me that this was something that made an impression on her. Perhaps in the future, when she enjoys the blessings of the Lord, she will remember to respond with praise for her Lord and King. Is this not what really matters?

Ultimately, it should not matter how we prefer to worship the Lord. Being His servant requires that we praise and worship Him the way that He wants to be worshipped. And the Lord favors those who worship Him wholeheartedly. For the person with a more expressive personality in worship, that means you need to make sure that you are expressing yourself for His glory and not to bring attention to yourself. For the person like me, who is more conservative in worship, that means making yourself available to be more expressive with our Lord as He leads. We cannot worship the Lord wholeheartedly while refusing to raise our hands, clap or shout because it makes us uncomfortable. A joyful heart results in joyous praise. Ever since that first field trip praise session, I conclude each field trip with a time of appreciation for our Lord for all He has done. I believe that it is critical to establish this example for our children and students so that they can learn to incorporate worship into all that they do and live lives that are pleasing to Him.

Personal Reflection & Application

1. What has the Lord revealed to you by reading this section?

2. How can you worship the Lord more fully with your children in your science classes?

3. What is your “preferred” style of worship? Are you allowing your pride to hold you back from worshipping the Lord more fully?

Teach with Eternal Focus

Establishing a biblical foundation for your science classes will ensure that you prepare your children, not just for their earthly future, but also their eternal future. If you only focus on teaching about the natural environment, your lessons will poorly prepare them for what really matters because our world is only temporary. While temporary things will pass away, those who believe in Jesus Christ will be saved and enjoy eternal life through Him (John 3:16; 1 John 5:11). The temporary nature of our earth is confirmed in the Revelation of Jesus Christ.

Now I saw a new heaven and a new earth, for the first heaven and the first earth had passed away. Also there was no more sea.

– Revelation 21:1

This verse describes a dramatic and monumental change in our natural environment. Of specific interest to me as a marine biologist are those last six words. I had been a marine biologist for several years when I finally realized the impact of this verse. I was immediately filled with disappointment since my visions of heaven included spending eternity swimming underwater without need of SCUBA equipment or a submersible.

"Well, what on earth am I doing?" I thought. "Why would God give me a desire to be a marine biologist when it is only a temporary position?"

Non-biblical curriculum is without purpose. Teach your children with eternity in mind.

At that moment, the Lord started laying the foundation in my heart for this workbook. He started to reveal to me the importance of focusing on the intent of the work rather than the subject of the work. Regardless of the subject content, the intent must always be to glorify the Lord and elevate Jesus Christ or otherwise the work is pointless. As I continued reading, I also noted verse twenty-three which indicated that there would be no more sun or moon.

The city had no need of the sun or of the moon to shine in it, for the glory of God illuminated it. The Lamb is its light.

– Revelation 21:23

As I fought through the disappointment that these revelations brought, the Lord blessed me by making a correlation for me between these verses and Jesus Christ as our intermediary. He showed me that this earth is filled with intermediate forms that point to Jesus Christ's role as the intermediary for our sins. While the Lord ultimately sustains the earth and all that is in it, in our current environment, He uses light and water as intermediate forms to give life to His Creation. In our ecological system, there is nothing that can exist without water. It is the same with light. There are only a few organisms that can exist in environments without any light source. Both light and water are intermediate forms that God uses in our world to provide us with life. In the same way, Jesus Christ is our intermediary. The Bible says that sin separates us from God (Isaiah 59:2). Furthermore, "the wages of sin is death but the gift of God is eternal life in Christ Jesus our Lord" (Romans 6:23). Our sin prevents us from having access to the Father without a sacrifice to atone for our sins. While we have all earned death through

In eternity, God will be our direct Source of life. Jesus Christ has eliminated the need for anything that comes between us and direct access to our Father.

sin, Jesus Christ intercedes for us as the life giver by taking our place on the cross. Jesus Christ fulfilled our need for an intermediary by dying on the cross as a penalty for our sins and granting eternal life to those who believe in Him. His sacrifice again allows us to have access to God. This was symbolized at the moment of Jesus' death by the curtain in the temple that separated the holiest of holies from the rest of the building being ripped from top to bottom (Matthew 27:51). Jesus Christ, our intermediary, offers life, just like God uses water and light as intermediate forms to produce life in our world.

Since water and light are merely aspects of Creation that God designed to point to Jesus Christ, they are no longer needed in eternity. Instead, God will be our direct Source of life. The Lord's role as life-giver is symbolized in the new Heaven and the new Earth by the River of Life flowing directly from the Throne of God (Revelation 22:1). This is the only water mentioned as being present in our new environment. In addition, Jesus Christ will also act as the source of light (Revelation 21:23). Because of Jesus Christ's sacrifice, we will no longer be kept at a distance but instead be offered direct access to God. The Word of God indicates that if you only teach your children about the existing earth, they will only understand a shadow of things which will exist in heaven (Hebrews 8:5). You must focus on the eternal as well as the temporary.

Personal Reflection & Application

1. What has the Lord revealed to you by reading this section?

2. Why is it important for you to teach with eternity in mind?

3. What adjustments are you going to make to your science class to ensure that you are teaching with an eternal focus?

Enjoying God's Blessings

Teaching biblically-based science is incredibly freeing. This is because when we seek to honor the Lord in our classes, we place the outcome in His hands and we are free to enjoy His blessings. The Lord has repeatedly shown me during my years of teaching that that there are few things about my science classes that I can control, especially with field trips. Regardless of how carefully I prepare a lesson or an event, I am still unable to control the weather, what wildlife my students see, who attends or how many students are present. I cannot control if my computer crashes, if the projector starts displaying my slides in disco format or if the electricity goes out. But all of these things are in the Lord's control and He works all things together for the good of those who love Him (Romans 8:28). When we acknowledge Him and establish Him as the foundation for our classes, He loves to offer us abundant blessings.

One of the ways you can begin enjoying these blessings is by inviting the Lord to preside over your science classes as a banner. In doing so, you will position yourself and your children to receive much favor from the Lord. This concept is based on one of the lesser known names of the Lord, *Jehovah Nissi*, the Lord is My Banner.

The Lord has abundant blessings for those who come in His name.

And Moses built an altar and called its name, "The-Lord-Is-My-Banner"

– Exodus 17:15

I find it interesting that this name of God is used only once in the Bible, in Exodus 17:15. For context, I have included the entire account of the Israelites' battle against Amalek:

And Moses said to Joshua, "Choose us some men and go out, fight with Amalek. Tomorrow I will stand on the top of the hill with the rod of God in my hand." So Joshua did as Moses said to him, and fought with Amalek. And Moses, Aaron, and Hur went up to the top of the hill. And so it was, when Moses held up his hand, that Israel prevailed; and when he let down his hand, Amalek prevailed. But Moses' hands became heavy; so they took a stone and put it under him, and he sat on it. And Aaron and Hur supported his hands, one on one side, and the other on the other side; and his hands were steady until the going down of the sun. So Joshua defeated Amalek and his people with the edge of the sword. Then the Lord said to Moses, "Write this for a memorial in the book and recount it in the hearing of Joshua, that I will utterly blot out the remembrance of Amalek from under heaven." And Moses built an altar and called its name, The-Lord-Is-My-Banner: for he said, "Because the Lord has sworn: the Lord will have war with Amalek from generation to generation."

– Exodus 17:9-16

This passage reveals several truths about the character of our God. First, a banner symbolizes protection. When we place the Lord as our banner in our lives, He offers us protection, solace and comfort. Second, a banner announces our allegiance. When the Lord is our banner, we are announcing that we are allied with Him and we come in His name. The Word of God says that He blesses His children abundantly. By declaring our allegiance to the Lord and His Word, showing that allegiance by living lives of obedience and bringing glory to His great name, we place ourselves under His protection. We make ourselves available to receive His great blessings. And finally, when we place the Lord as our banner in our lives, we are establishing our priorities. We are saying that the Lord comes first over all else, that He sits on the throne to reign over us and that we serve Him. When we make the Lord our protection, our allegiance and our priority, we recognize His control and sovereignty and through Him, and Him alone, we will enjoy victory and His blessing, just like the Israelites. Truly, "If God is for us, who can be against us?" (Romans 8:31b).

Practically speaking, you can apply this concept to your science class by initiating each class session with prayer. In my opening prayers that I offer at the beginning of my classes and field trips, I ask for seven specific things from the Lord:

- His favor and blessings,
- That He will reveal Himself to us through His Creation so that we can grow in our knowledge of Him,
- That He will show my students how He blesses those who come in His name,
- That He will open the eyes of my students so they will be able to see His character in what He has made,
- That my students will enjoy real relationships with Jesus Christ rather than being trapped in patterns of religiosity,
- That I will teach what He wants me to teach, and
- That He will be glorified.

In sharing this list, my intention is not to provide you with a formula for prayer but rather to offer suggestions as to what you could include in your opening prayer during your science lessons with your children. As each of us is unique, I encourage you to pray as you are led by the Holy Spirit. And finally, at the end of your classes, please remember to offer praise and thanksgiving to the Lord for all that He has shown you during your class time or field trip. In following this approach, the Lord has shown me that He has abundant blessings for those who come in His name.

Let me give you an example by sharing with you one of my best experiences on a field trip. I was leading a relatively large group of students in a beach survey with a simple plan to climb a pile of boulders and search for marine life. These boulders were installed as a wave break on the beach and had become home to a wide variety of different sea creatures. On this particular day, when we arrived at the field trip location I was surprised to see an abundance of seaweed and wrack all over the beach. My students and I were excited to dig through the piles of seaweed to see what treasures we could find. There was an abundance of colorful sea shells, elaborate sponges, delicate algae, spiny sea urchins, flower-like corals, wavy soft corals and tropical sea beans everywhere. The volume of marine life present on the beach was astonishing!

While we were walking along the beach enjoying all that we were seeing, I also found out that the fishing jetty had reopened that same week after being closed for six years. After spending some time finding exciting specimens in the seaweed, I led my students down the newly renovated jetty. The jetty in question is along the south side of the Port Everglades inlet so there is a lot of boat traffic that passes by on the way in and out of the Port. While we were walking on the jetty we saw two 80-foot tugboats leaving the Port to escort a tanker moored offshore. My students were so excited to see these large vessels that they started shouting and waving at the boats. In response, one of the boat captains stopped directly in front of my students, turned a full circle and sounded his horn for them. My students went crazy, clapping, jumping and shouting. As we continued on the jetty, the fishermen on the jetty enthusiastically displayed their catch to my students, allowing them to touch and hold them. Reaching the end of the jetty, we paused to observe a pair of parrotfish nibble algae off of the rocks. As we watched their brightly colored bodies flash in the sun, one of my students next to me yelled, "Hey look! A manatee!"

Right in front of us, a manatee slowly glided past, surfaced and flipped its tail, hanging around just long enough for all of my students to see him. My students cheered.

On the way back to the parking lot, one of my students turned to me with a huge grin on her face, "This was the best field trip ever!"

I smiled, "We need to praise the Lord for His blessings today because I certainly didn't have anything to do with it."

She laughingly responded, "Miss Christa, you mean that you didn't have that manatee on remote control?"

I laughed with her, "No, I did not. But the Lord sure did."

While this might seem like a fitting closing moment for a near perfect day, the Lord had one final surprise in store for us. Right before we left the jetty, the Blue Angel fighter jets soared past on one of their practice runs for the Air Show that was scheduled for the weekend. We later found out that this was one of the few times they were seen that weekend because of poor weather conditions. The Lord had sent them out for our enjoyment. How great is our God?

I could fill the rest of this workbook with numerous other stories of the Lord's abundant blessings that I have enjoyed through teaching my science classes. But instead I challenge you to establish the Lord as your banner over your science class so that you can see the blessings that He has in store for you and your children. What will the Lord, in His great faithfulness and mercy, reveal to you about His character and His Creation?

Personal Reflection & Application

1. What has the Lord revealed to you by reading this section?

2. How can you make the Lord your "banner" in your science classes? In your daily life? In your relationships?

3. How can you teach your children to make the Lord a "banner" in their lives?

Introducing Biblical Environmentalism

In addition to teaching a biblical perspective on what exists in the world, a biblically-based science class must also give the appropriate perspective on how we are to take care of it. You must also teach your children about biblical environmentalism. Based on the world's interpretation of environmentalism, it may seem like these two words contradict each other. However, since the Lord made all of Creation, we should seek His instructions on how to take care of it. Quite simply, biblical environmentalism is learning how to take care of Creation and all that the Lord has made in a manner that is pleasing to Him.

In underscoring the importance of reviewing the Bible and what it says about caring for Creation, I am going to highlight a few critical points. First, the Lord values all of Creation as it reflects His character and beauty.

> The heavens declare the glory of God; the skies proclaim the work of his hands. Day after day they pour forth speech; night after night they reveal knowledge.
>
> – Psalm 19:1-2 NIV

A few months ago I saw a brief clip of a session by Louie Giglio from the 2011 Desiring God National Conference. In his message he played clips of the sounds made by oscillating stars in the heavens. Just like Psalm 19 says, these stars pour forth "speech" for twenty-four hours a day, 365 days a year. In doing so, they are declaring the glory of the Lord and illustrating the truth of God's Word.

Second, God's Word is also very clear that we do not own the earth. The earth, its resources and all the animals and plants that live on it are owned by the Lord.

Biblical environmentalism is learning how to take care of Creation and all that the Lord has made in a manner that is pleasing to the Lord.

The earth is the LORD's, and everything in it, the world, and all who live in it; for he founded it upon the seas and established it upon the waters.

– Psalm 24:1-2 NIV

We also see God's ownership displayed in the system of government He established for the Israelites. The Lord gave each tribe and family of Israel land as their inheritance. In this agrarian society, land represented the opportunity for a family to sustain themselves by cultivating crops and raising livestock to meet their needs. In the event that the Israelites found themselves in debt for whatever reason, they could sell their land but in doing so, they would limit their opportunities for financial prosperity in the future. In order to ensure that a family would not lose their inheritance forever the Lord established the procedure for the year of Jubilee in which any land that had been bought or sold would revert back to its original owner every fifty years (Leviticus 25).

The land must not be sold permanently, because the land is mine and you reside in my land as foreigners and strangers.

– Leviticus 25:23 NIV

While this law shows the mercy of the Lord, it also shows that He wanted to develop a culture of stewardship, not ownership, in the Israelites.

Third and finally, the Bible also clearly states our role in regards to the earth. God's Word indicates that we are the caretakers of all Creation. This means that we have dominion over the earth, its plants and animals. We are to manage it and use its resources to meet our needs but not to abuse it. We must take care of the earth as it is a gift that has been entrusted to us by the Lord. There are numerous passages in the Bible that illustrate these concepts.

God blessed them and said to them, "Be fruitful and increase in number; fill the earth and subdue it. Rule over the fish in the sea and the birds in the sky and over every living creature that moves on the ground."

– Genesis 1:28 NIV

The Lord God took the man and put him in the Garden of Eden to work it and take care of it.

– Genesis 2:15 NIV

For the Lord your God is bringing you into a good land—a land with brooks, streams, and deep springs gushing out into the valleys and hills; a land with wheat and barley, vines and fig trees, pomegranates, olive oil and honey; a land where bread will not be scarce and you will lack nothing; a land where the rocks are iron and you can dig copper out of the hills.

– Deuteronomy 8:7-9 NIV

The highest heavens belong to the Lord, but the earth he has given to mankind.

– Psalm 115:16 NIV

These passages of Scripture model for us the appropriate way that we are to interact with the Lord's Creation for the purpose of bringing honor and glory to Him. I encourage you to begin instilling these values in your children by mentioning them in your science class. It is important to teach them the truth about the environment since there are so many different opinions about the world and our appropriate relationship with it. By knowing what God says in His Word, your children will be able to correctly discern the truth.

Personal Reflection & Application

1. What has the Lord revealed to you by reading this section?

2. What is biblical environmentalism?

3. What is our role in relationship to God's Creation and how are we to interact with it?

Encouraging Good Stewardship

As a follower of Jesus Christ, it is imperative that we maintain the right balance between conserving and protecting the earth while also using the resources that the Lord gave us to meet our needs. This type of balance is not something that we often see in our world today. It seems that people generally gravitate to one of two extremes: over protection or over utilization.

Representing one extreme are the environmentalist groups who focus on protecting the earth by placing preservation of natural resources and animals above the needs of humans. This strategy is contrary to the Word of God. Human beings are the only creature that the Lord created in His image. Unlike any other creature on the planet, He gave us an eternal soul and He sent His Son, Jesus Christ to die on the cross for *us*. Animals are not our equals. By placing the welfare of animals at a predominant level of importance, we see that something intended to be a blessing from the Lord has become an idol and a stumbling block. Rather than believing the truth of God's Word, we see the people of this world falling for a lie.

We must maintain the right balance between conserving and protecting the earth while also using the resources that the Lord gave us to meet our needs.

They exchanged the truth about God for a lie, and worshiped and served created things rather than the Creator—who is forever praised. Amen.

– Romans 1:25 NIV

As discussed in Romans 1:25, by falling prey to this lie, people are choosing to worship created things rather than the Creator Himself. Consequently, we see strategies employed like those of a group of scientists in Canada who are seeking support for a Declaration of Rights for Cetaceans. The purpose of this petition is to extend human rights to dolphins and whales due to their high levels of intelligence. By doing so, the authors are intending to use it as a way to end whaling, the use of dolphins in entertainment and their captivity.[3] While on the surface these may seem like noble goals, once dolphins are granted the same rights as humans, where will their rights end? Will they be able to own property? Get married? Will children be able to choose dolphin as a foreign language in school? The scenarios become ridiculous rather quickly.

Representing another extreme on the opposite end of the pendulum from the conservationists are lovers of money who exploit the natural environment to gain wealth. Their greed is in itself another form of worship. However, rather than being worshippers of created things, these people are worshipping at the altar of fortune. Jesus Christ warns against this type of service in the Sermon on the Mount.

> No one can serve two masters. Either you will hate the one and love the other, or you will be devoted to the one and despise the other. You cannot serve both God and money.
>
> – Matthew 6:24 NIV

According to the Bible, we must decide for ourselves whom we serve because we cannot serve the Lord wholeheartedly while also trying to serve ourselves by amassing personal wealth. Unfortunately, this type of misplaced worship results in the total exploitation of the Lord's Creation, in direct disobedience to the Word of God. One dramatic example we see is through the inhumane process of shark finning. Approximately 26 to 73 million sharks are killed *every year* through shark finning.[4] The practice of shark finning includes commercial fisherman catching a live shark and then removing its dorsal and pectoral fins without killing it. The shark is tossed back into the water to suffer and die slowly. Once their fins are removed, sharks are no longer the beautifully aerodynamic and graceful swimmers God created. Since they are not capable of keeping themselves upright without fins, they roll through the water and finally sink to the sea floor. Eventually finned sharks die from starvation, suffocation from the lack of oxygen being transported over their gills or are eaten by other marine predators. Sharks are finned in this manner because shark fin soup is a delicacy in Asian countries. Restaurants charge up to $100 per bowl of shark fin soup. For illustrative purposes, let us assume that fins from one shark result in about 10 bowls of soup. This means that approximately $1,000 of income is represented by each shark. Assuming this is the case, the shark fin industry then generates annual gross revenue of $26 to 73 billion dollars. What a gross waste and desecration of the Lord's Creation in the name of greed.

It is so critical that we as believers in Jesus Christ do not follow the patterns of this world in their worship or exploitation of natural resources. We must model the appropriate balanced approach to Creation which involves enjoying the earth but not worshipping or exploiting it. We must use the natural resources that the Lord has given us as needed without being wasteful. In this way, we will honor the Lord and the gift that He has given us. It is important to emphasize biblical environmentalism in your classes so that your children will be able to discern the appropriate way to interact with Creation in a manner that glorifies the Lord.

Personal Reflection & Application

1. What has the Lord revealed to you by reading this section?

2. What are two extremes we see in today's society regarding human interaction with Creation?

3. How are followers of Jesus Christ called to act differently?

A God Who is for Us

As a conclusion to this portion of the workbook, I would like to offer an important reminder that our God is faithful all the time, regardless of our current circumstances. This is a critical point to remember because there will be times in which you will face opposition from others in regards to teaching a science class based on the foundation of God's Word. As you seek to honor the Lord in your science classes, you will be reminded that developing a biblically-based approach to science is not popular in today's culture. Consequently, you may find that you face opposition from other people both spiritually and relationally. Non-believers may criticize, ridicule and try to intimidate you. I have had instances during my field trips when I have been approached by people on the beach who felt the need to disagree with my lessons. Each of these interactions was a great way for me to demonstrate to my students how to respond to these individuals with truth and love. Sadly, you may also face opposition from those in the church who have compromised their position on a literal interpretation of the Bible. In these moments of difficulty, I encourage you to remember that establishing a biblical foundation for your science classes requires a spirit of boldness that is consistent with living a life dedicated to serving the Lord.

But we do not belong to those who shrink back and are destroyed, but of those who have faith and are saved.

– Hebrews 10:39 NIV

And let us not grow weary while doing good, for in due season we shall reap if we do not lose heart.

– Galatians 6:9

The Bible has repeated statements of exhortation reminding us that those who persevere will ultimately be rewarded by the Lord. God also tells us that He has granted us a spirit of power, of love and a sound mind (2 Timothy 1:7). With these tools offered through the Holy Spirit, we are equipped to handle any attack thrown our way.

Finally, I also want to remind you that our God is for us. The Word of God says that He is looking for opportunities to show Himself on our behalf.

For the eyes of the Lord run to and fro throughout the whole earth, to show Himself strong on behalf of those whose heart is loyal to Him.

– 2 Chronicles 16:9a

With the God of the Universe on our side, we can be more than conquerors through Jesus Christ. I encourage you to remember these valuable truths when you are facing opposition, discouragement or doubt. When you prioritize the Lord and seek to honor Him in your science classes, He will be your defender and the lifter of your head.

Personal Reflection & Application

1. What has the Lord revealed to you by reading this section?

2. How have you faced opposition from others in regards to your science classes or homeschooling in general? Spend some time in prayer asking the Lord to help you to face such opposition with boldness while presenting the truth in love.

3. Which of the verses mentioned in this section was most meaningful to you and why?

PART 2

Addressing the Creation Versus Evolution Issue

A Polarizing Debate

In the beginning God created the heavens and the earth.

– Genesis 1:1

Evolutionary theory is one of the most prevalent issues in our society. Debates between evolutionists and Creationists lead to much controversy and confusion. Many times it seems that both sides of the argument are using the same data to come up with totally opposite conclusions. In order to assist you in addressing these issues with your children in science class, this portion of the workbook will discuss a biblical perspective on this hotly contested topic.

Have you ever wondered why the Creation versus evolution debate is so polarizing? It is because evolutionary theory has nothing to do with science. Evolution is a spiritual, not a scientific, issue. In order to be scientific, a theory must be able to be tested and the results replicated for conclusions to be drawn. Evolutionary theory is unable to be tested because at its base is the requirement for change over millions of years, a process that cannot be duplicated or observed. Instead, evolution provides men with an opportunity to eliminate God from the start of the world thereby absolving themselves from the moral absolutes established by Him in the Bible.

The fool has said in his heart, “There is no God.”

– Psalm 14:1a

The concept of this world coming into existence through natural processes, with matter slowly evolving from simple to complex organisms, cannot even be considered unless it is first assumed that there is no God. Evolutionary theory is nothing more than a satanic attack on the Word of God dressed up to look like science.

Evolution is a spiritual, not a scientific, issue.

This truth allows us to confidently look to the Word of God for our appropriate response to evolutionary theory. Regardless of the intimidating words of scientific heavyweights like Sir Peter Medawar, who said, "The alternative to thinking in evolutionary terms is not to think at all,"[1] a scientific background is not needed to consider evolutionary theory. All that is required is a heart that desires to hear from the Lord Jesus and to seek the truth in His Word.

Since evolution is a spiritual issue rather than a scientific one, we need to evaluate it by using the appropriate biblical test.

"Beware of false prophets, who come to you in sheep's clothing, but inwardly they are ravenous wolves. You will know them by their fruits. Do men gather grapes from thornbushes or figs from thistles? Even so, every good tree bears good fruit, but a bad tree bears bad fruit. A good tree cannot bear bad fruit, nor can a bad tree bear good fruit. Every tree that does not bear good fruit is cut down and thrown into the fire. Therefore by their fruits you will know them.

- Matthew 7:15-20

The Word of God says that we can discern false teaching by evaluating the fruit it generates. As we apply this biblical test to evolution, it becomes evident that evolutionary theory is not from the Lord. When we look at the fruit produced by evolution-based thought, we clearly see the mark of Satan. The Bible says that Satan is the father of lies and a murderer (John 8:44). Satan's purpose in this world is to steal, kill and destroy while Jesus Christ offers abundant life through Himself (John 10:10). If we look into the history of evolution and its impact on the human race, we can clearly see the swath of destruction it has left in its wake.

In his book *Evolution's Fatal Fruit*, Tom DeRosa sheds light on the links between Charles Darwin's evolutionary theory outlined in his book *Origin of Species* and Adolf Hitler's policy of Caucasian racial superiority and ethnic cleansing.

> "Darwin's ideas made Nazi atrocities possible. Anti-Semitism had existed in Europe for centuries before Hitler, but Darwin provided the intellectual catalyst for action. The Nazi program would not have been possible without the ideas of 'natural selection and the preservation of favoured races in the struggle for life' – Darwin's subtitle for *Origin*. These ideas did not originate with Hitler. They originated in Darwin's notebook when he drew his tree of life and theorized that good would come from death via natural selection."[2]

The attempted extermination of the Lord's chosen people can be directly linked to evolutionary thought.

Unfortunately, the satanic forces behind the atrocities of the Holocaust continue to manipulate and deceive people today through the theory of evolution. In performing research for this workbook, I came across a quote attributed to David Quammen, an American science writer. It said, "Hope is a duty from which paleontologists are exempt." This is a desperate statement cloaked in a mask of indifference. As I researched this quote, I was horrified when I came across the original article titled "Planet of Weeds: Tallying the losses of Earth's animals and plants". The article was posted on the Church of Euthanasia's web site. This is an actual non-profit organization legally registered as a corporation in Massachusetts. According to the mission statement posted on their web site, they are "devoted to restoring balance between Humans and the remaining species on Earth. We believe this can only be accomplished by a massive *voluntary* population reduction, which will require a leap in Human consciousness to a new *species awareness*."[3] The rest of their web site includes voluminous content glorifying suicide, abortion and murder based on the justification that the elimination of large numbers of

Rather than recreating the Creation versus evolution debate, this workbook will focus more on the principles behind biblically-based teaching and the Creation versus evolution issue. I recommend that you seek out other resources that provide a complete scientific defense for Creationism in order to ensure that you have a well-balanced approach in your science class. While there are numerous wonderful ministries that tackle this critical issue, two that I highly recommend are: Answers in Genesis (www.answersingenesis.org) and the Creation Studies Institute (www.creationstudies.org).

the human race is the only way that our planet can survive. In their minds, humans are parasites, living and thriving at the earth's expense. What is most distressing about this organization is that the people involved have no concept of their intrinsic value to their Creator (Psalm 139). They do not believe that there is a loving God who has a purpose for their lives (Jeremiah 29:11) or that children are a blessing from the Lord (Psalm 127:3-5). What a desperate and hopeless existence! They are clearly under a satanic influence as they openly and unapologetically promote the indiscriminate destruction of the human race. Ladies, please pray for people such as these, that they will be rescued from their pit of despair and come to a saving knowledge of Jesus Christ. While most evolutionists are not nearly as extreme as those involved with this organization, ultimately, this is the fruit of their beliefs. The truth behind evolution is ugly. Christians are dishonoring their God by having anything to do with such blatant mockery of His Word. As a follower of Jesus Christ, the theory of evolution has no place in your science class.

Personal Reflection & Application

1. What has the Lord revealed to you by reading this section?

2. Is evolution a scientific theory or a spiritual issue? Why?

3. How can we discern that evolutionary theory is not from the Lord?

Spiritual Origins of Evolutionary Theory

Evolution is one of the greatest lies of the past few centuries. Since its inception, we have seen an erosion of the moral foundations of our society as our children are taught to question biblical truth, the creation of the world and the purpose of their very existence, all in the name of "science". Once again, we do not see any evidence of God in this theory, but we see the fingerprints of the great usurper himself, Satan. Here is what the Bible says about Satan's character:

Just like we can see the character of God in what He has made, we can also see the influence of Satan in evolutionary theory.

- He was first created as a beautiful creature, Lucifer, who was designed to worship and praise the Lord (Ezekiel 28:13).
- His sin rooted in his prideful and violent heart caused him to be thrown out of heaven (Ezekiel 28:16-17).
- He desires to be worshipped as a god (Ezekiel 28).
- He is the father of lies (John 8:44).
- His intent is to kill and destroy (John 10:10; 1 Peter 5:8).

Just like we can see the character of God in what He has made, we can also see the influence of Satan in evolutionary theory.

"Your heart was lifted up because of your beauty; you corrupted your wisdom for the sake of your splendor; I cast you to the ground, I laid you before kings, that they might gaze at you."

– Ezekiel 28:17

In comparison, let us consider the beliefs at base of evolutionary theory:

- The theory assumes that there is no divine Creator but instead earth was created through natural processes.
- Since there is no God, evolutionists are freed from the moral absolutes of the Bible.
- Worship is misdirected towards the creation itself rather than the Creator. We often see this manifested as worship of self or by placing the importance of animals and plants above the needs of humans.
- Evolution encourages pride in the hearts of men as they seek to provide rational answers for all of life's mysteries.

Evolution is just a new name for an old game. If we review the content of the Bible, we can see Satan, the great deceiver, using his influence to cause people to question the Lord and seek to take His place in their lives. We see this pattern in two major events in the Bible that influence the entire course of history: the fall in the Garden of Eden and the chaos at the Tower of Babel.

Now the serpent was more cunning than any beast of the field which the Lord God had made. And he said to the woman, "Has God indeed said, 'You shall not eat of every tree of the garden'?" And the woman said to the serpent, "We may eat the fruit of the trees of the garden; but of the fruit of the tree which is in the midst of the garden, God has said, 'You shall not eat it, nor shall you touch it, lest you die.'" Then the serpent said to the woman, "You will not surely die. For God knows that in the day you eat of it your eyes will be opened, and you will be like God, knowing good and evil." So when the woman saw that the tree was good for food, that it was pleasant to the eyes, and a tree desirable to make one wise, she took of its fruit and ate. She also gave to her husband with her, and he ate.

– Genesis 3:1-6

If we are not careful, it could be easy to think that the original sin was simply an issue of Eve's disobedient heart. A closer look reveals something deeper in Satan's temptation. He offers Eve the opportunity to *be like God* and she immediately responds. Her desire was to share the wisdom of the Holy One, making herself equal with Him. We see the same pattern in the account of the Tower of Babel.

Now the whole earth had one language and one speech. And it came to pass, as they journeyed from the east, that they found a plain in the land of Shinar, and they dwelt there. Then they said to one another, "Come, let us make bricks and bake them thoroughly." They had brick for stone, and they had asphalt for mortar. And they said, "Come, let us build ourselves a city, and a tower whose top is in the heavens; let us make a name for ourselves, lest we be scattered abroad over the face of the whole earth."

– Genesis 11:1-4

Again, the desire of the people in building the Tower of Babel was to become equal with God. They wanted to make a name for themselves, not their God, by building a tower that reached His dwelling place in the heavens. My personal belief is that somehow all sins ultimately return to this issue of pride. While they may manifest in different ways, at the root of every sinful, anti-God decision is the spirit of pride.

This is why we are warned in the first chapter of James that ultimately our own desires are what lead us to temptation and death.

> But each one is tempted when he is drawn away by his own desires and enticed. Then, when desire has conceived, it gives birth to sin; and sin, when it is full-grown, brings forth death.
>
> – James 1:14-15

Only by crucifying our sinful nature, our passions and our desires, thereby submitting to the only true God, are we able to serve Him (Galatians 5:24). As we examine these passages of scripture, we can see clearly once again that evolutionary theory is a direct attack on the Bible and God Himself. As followers of Jesus Christ, we must be able to discern these attacks on Scripture and train our children to perceive them as well. Simply eliminating evolution from our lessons is not sufficient. We must discuss both the scientific evidence and the Biblical truth that disprove evolutionary theory so that our children will be prepared to respond appropriately to these attacks.

Personal Reflection & Application

1. What has the Lord revealed to you by reading this section?

2. How does evolutionary theory reflect the character of Satan?

3. What similarities can we see between evolutionary theory and the biblical accounts of Adam and Eve's original sin and the Tower of Babel?

Separating Light from Darkness

By examining the fruit and spiritual foundation of evolutionary theory, we can recognize that evolution is clearly not of the Lord. For this reason, it is imperative that we maintain a pure perspective of the Creation account that does not incorporate any of the facets of evolution. Unfortunately, we see people today who claim to be believers in Jesus Christ yet have been swayed by seemingly scientific information to think that evolution is true. This propaganda starts to chip away at their confidence in the Bible by causing them to view scriptures out of context so that they can mold them to match the current culture.

One of the most frequent challenges to the Creation account that I hear from Christians generally sounds something like this: "Perhaps a day in the Creation account is not literally a twenty-four hour period. After all, the Bible says a day is like a thousand years to the Lord."

For additional information on historical and scientific research that supports the Bible, you can visit web sites such as Answers in Genesis (www.answersingenesis.org) or Creation Studies Institute (www.creationstudies.org).

Unfortunately, this statement is based on an incorrect interpretation of 2 Peter 3:8 which says, "with the Lord one day is as a thousand years, and a thousand years as one day." The purpose of this scripture is simply to indicate that God is not limited by time and in His great mercy has delayed His return to allow the people of the earth to repent to avoid eternity in the Lake of Fire. However, for the sake of addressing this statement, if we use this verse to interpret every day in the Creation story as one thousand years, then the reverse must also be true. Every time that scripture mentions one thousand years, we must conclude that it is actually referring to one single day. However, historical and archaeological evidence easily disproves this conclusion.

In addition, our Lord God, in His infinite wisdom, already addressed this challenge to the Scriptures before it was even presented. A closer look at the Creation account in the first chapter of Genesis reveals that the Lord specifically defines a Creation day as a twenty-four hour period. At the conclusion of the account of what was created on each day, the Bible finishes with a definitive statement saying that there was "evening and morning, the 'x' day" (Genesis 1:5, 8, 13, 19, 23, and 31). Is it not interesting that God chose to mention the specific time span for each Creation day after every single one of the six days? By repeating this phrase, He is essentially underscoring the fact that the earth was created through spoken word in just six short twenty-four hour periods. This interpretation is confirmed later in the Bible as well.

For in six days the Lord made the heavens and the earth, the sea, and all that is in them, and rested the seventh day. Therefore the Lord blessed the Sabbath day and hallowed it.

– Exodus 20:11

Consequently, the Bible does not allow any room for evolution in the Creation account.

In addition, as believers of Jesus Christ who seek to honor Him, we must not become associated with thought patterns, beliefs or actions such as evolutionary theory that are dishonoring to Him.

Do not be unequally yoked together with unbelievers. For what fellowship has righteousness with lawlessness? And what communion has light with darkness?

– 2 Corinthians 6:14

We are called to be set apart from the world. More importantly, we are called to be holy because He is holy (1 Peter 1:16). The Lord does not tolerate compromise, even saying that lukewarm beliefs are disgusting to Him (Revelation 3:16). By allowing ourselves to become influenced by evolutionary theory, we are giving up the opportunity to be a light to unbelievers in this world. It is our role to draw men to Jesus Christ rather than allowing men to draw us into sin. Accepting evolution is an attempt to accommodate current culture that requires changing the interpretation of the Bible. You must use your science classes as a way to encourage your children to maintain their own Biblical purity so that they can be the light of Jesus Christ in our dying world.

Personal Reflection & Application

1. What has the Lord revealed to you by reading this section?

2. Can the Creation account be interpreted to include evolutionary theory? Why or why not?

3. Why is it important for followers of Jesus Christ to maintain a pure perspective of the Bible?

Capture Your Child's Heart with Truth

If considered at purely a face value, evolutionary theory sounds ridiculous and totally illogical. My intent is not to be condescending to those who believe in evolutionary theory but rather to bring light to the truth. Evolution is based on the belief that simple organisms like amoebas evolved through genetic mutations into the complex organisms we see today. This means that both plants and animals could possibly share the same amoeboid ancestor billions of years back in the family tree. By just extrapolating this thought a little further, you could possibly conclude that you and the tree in your front yard are related. As you follow the pattern of evolutionary thought and all its implications, you can see why people in this world who have fallen prey to this lie devalue human life and place such great importance on the preservation of "Mother Earth". This is because in their minds, the earth and all its inhabitants really are their relatives. It sounds more than a little far-fetched, right? Without understanding the spiritual aspects behind evolutionary theory, it is hard to believe that so many people in this world would believe such foolishness. Unfortunately, people who are lost in their sin easily fall prey to the lies of Satan and are spiritually blind to the truth. Without the Holy Spirit to open their eyes, they do not realize how deceived they truly are. Knowing the truth through the power of the Holy Spirit gives us two major responsibilities: to pray for those who are deceived and to train our children to discern the truth. Consequently, you must seek to capture your children's heart at a young age with the truth of God's Word before they fall prey to Satan's lies.

What do you think your children would say if during one of your science classes, you took them outside to a tree or bush and introduced it to them as one of their extended family members? Can you hear the laughs and giggles from your youngest ones? Or maybe feel the disbelieving stares from your older children that seem to say, "Mom's really lost it this time"? Simply put, kids get it. This is because their hearts are able to see the truth before pride can take root in their lives and give them the desire to eliminate the need for God. The Bible confirms this truth in Matthew 18:3-5 where Jesus Christ states that only those who are as humble as little children will be saved.

Religious behavior builds division but relationship creates unbreakable bonds.

Then Jesus called a little child to Him, set him in the midst of them, and said, "Assuredly, I say to you, unless you are converted and become as little children, you will by no means enter the kingdom of heaven. Therefore whoever humbles himself as this little child is the greatest in the kingdom of heaven. Whoever receives one little child like this in My name receives Me.

– Matthew 18:2-5

In order to prevent children from falling prey to the deception of evolution, we must train them to be humble and seek to know the Lord through His Creation. It is only by the grace of God and through the truth of His Word that they will be able to reject the advances of the enemy when they have matured into adulthood. By using your science classes to train them to see the Lord in all that He has made they will automatically know to reject the theory of evolution and its lies. It will not even be a consideration for them because their God has become a personal God that they intimately know and have experienced rather than one they heard about at church. Religious behavior builds division but relationship creates unbreakable bonds. Use your science classes to help your children build a transforming relationship with Jesus Christ that will sustain them throughout their entire lives.

Personal Reflection & Application

1. What has the Lord revealed to you by reading this section?

2. Why have people today fallen for the lie of evolution?

3. Knowing the truth about evolution, what are our God-give responsibilities?

4. Why is it important to capture your child's heart with the truth at a young age?

PART 3

Practical Tips for Your Science Class

Your Kids Say Your Science Class is Boring. Now What?

Over the past year or so, I have written nearly 100 hours of class curriculum. There are some days when ideas flow and it seems easy for me to come up with new and creative ideas for my students. However, there are other days when I am lacking inspiration and curriculum writing seems like a stressful and immensely difficult task. During these moments, I have learned to pause and ask the Lord to grant me the ideas I need to do the work that He has called me to do. The Lord knows the hearts, talents and interests of my students and He has shown me that He is able to provide me with insights that will be effective tools in meeting their educational needs.

Learn to pause and ask the Lord to grant you the ideas you need to do the work to which He has called you. The Lord is faithful to provide for all of your needs.

There was a situation about a year ago, where one of my friends, also a homeschooling mom, asked me to do a hands-on class for her children at the beach. While this is a great idea, locally beaches in South Florida present some challenges. The beaches in the Fort Lauderdale area are maintained where beach cleaners remove both trash and natural debris from the waterline creating an area generally sparse in marine life. High numbers of visitors also result in fewer chances to see marine life as wild animals generally avoid populated areas. Consequently, there just is not much to see at the local beach. While I was initially hesitant, I really liked my friend's idea so I went to the beach with a notebook to pray for inspiration from the Lord. As I sat in the sand and looked around me, the Lord gave me the idea for a four-week hands-on Course that would teach students about beach erosion, waves and currents, biological communities and environmental sampling. In addition, the Lord provided me with a connection with a local environmental company that rents the equipment I needed for the sampling class. Because of the Lord's favor, they allow me to use $6,000 worth of sampling equipment for free for educational purposes. If we seek Him and trust Him, the Lord provides for every work that He has called us to do.

Even though I now have more experience in writing curriculum, I continue to seek the Lord for His direction and inspiration. I like that there are still times when this is a difficult process for me. If it was not, it would be very easy for me to fall into a trap of writing "good" curriculum rather than "God-inspired" curriculum. If we are seeking to teach a class that honors the Lord, is it not appropriate to ask Him what He wants us to teach?

As a final thought, I encourage you as you are preparing your own lessons to remember that God is the author of all creativity. He is a wellspring of information that you can never exhaust. He is able to give you every idea and resource that you need to offer an exciting and fun science class for your children that honors and glorifies Him. All you have to do is ask (James 4:3).

Personal Reflection & Application

1. What has the Lord revealed to you by reading this section?

2. What is the appropriate response when faced with a challenging or overwhelming situation?

Looking for God's Fingerprints

Our God is an interactive God. Consequently, our science classes should be interactive as well. God enjoys spending time, communicating and building relationships with His children. Since one of the main purposes of Creation is to bring us into a relationship with the Lord, a logical conclusion is that we need to spend time *in* Creation. Do not be afraid to take your science class outside. Oftentimes, a fun field day will teach your children more about Creation, scientific processes and our God, than the most productive day doing book work at home.

In addition to trips to the library, fire station or local athletic complex, I highly recommend spending time in the natural environment. Give your children the opportunity to experience Creation in a hands-on manner. Let them touch, see, feel, hear, taste. Creation is meant to be experienced so that God can also be experienced.

I also encourage you to create interactive experiences where your children need to search for the character of God in Creation. I implement this in my own classes by teaching a portion of the class and then asking my students questions such as:

- How can we see the character of God in what we just saw, learned about, discussed?
- Can you think of a scripture verse that would apply to that characteristic?
- How can you relate to that on a personal level?

You may want to begin by providing your children with a list of the characteristics of the Lord, then ask them to choose a character from the list that they see displayed in Creation and describe how it can be seen. This is even more fun to do in the natural environment where they can make observations of Creation and correlate those observations with the Lord.

It is also important for you to encourage them to think about what they have seen. They will probably need time to consider and digest what they have experienced. Perhaps you can incorporate a time of reflection into your lessons following your field trips and give them an assignment where they respond to your questions in a notebook or field journal. To assist you in including this important facet into your lessons, I have given detailed examples in the next section for your reference.

Do not be afraid to take your science class outside. Oftentimes, a fun field day will teach your children more about Creation, scientific processes and our God, than the most productive day doing book work at home.

In walking through these exercises with my students, I have discovered that this is a learning process. The first time I asked a student to tell me what character of God they could see in a marine ecosystem, all I received in response was a blank stare. But over time, they learned to correlate the truth of His Word and the characteristics of God they saw listed in Scripture with the world they observed. And as you and your children seek to know the Lord through His Creation, He will be faithful to continue to reveal more and more of His character to you. As the Word of God says, "You will seek Me and find Me, when you search for Me with all your heart" (Jeremiah 29:13). God is waiting to reveal Himself to you. All you have to do is look for Him in the world around you.

Personal Reflection & Application

1. What has the Lord revealed to you by reading this section?

2. Why is it important to spend time in the natural environment on field trips with your children?

Attributes of God

When I am teaching students to look for God's characteristics in Creation, I often give them a list of the various characteristics of the Lord and ask them how we can see those characteristics in what He has made. This process takes training because for many students this is a new way of thinking. But once they learn to consider what they are seeing in Creation, God will continue to use this process to reveal Himself to them through what He has made. To facilitate these types of activities with your children, I have included a list of attributes for your reference. While this list is not all-encompassing, it will help you get started in using this approach in your science class.

Teaching your children how to see the character of God in Creation takes training because for many students this is a new way of thinking.

- Awesome: 1 Chronicles 16:30; 2 Chronicles 7:1-3; Nehemiah 9:5-6; Isaiah 6:5
- Beautiful: 1 Chronicles 16:29; Psalm 27:4, 29:2
- Body of Christ: 1 Corinthians 12:12
- Eternal: Genesis 21:33; John 8:58; Revelation 1:8
- Faithful: Deuteronomy 7:9; Joshua 23:14; Psalm 100:5; 1 Corinthians 1:9; 1 Thessalonians 5:24
- Friend: Proverbs 18:24; John 15:15
- Good: Genesis 1:31; Psalm 119:68; Nahum 1:7; Mark 10:18
- Gracious: Exodus 34:6-7; Nehemiah 9:17; Psalm 103:8; 2 Timothy 1:9; Titus 3:5-7
- Great: Psalm 77:13, 145:3
- Help: Psalm 54:4; Hebrews 13:6
- Hiding Place: Psalm 32:7

- Holy: Isaiah 6:3, 57:15; 1 Peter 1:15-16; Revelation 4:8
- Incomprehensible (beyond human understanding): Job 36:26; Isaiah 40:18-26; Matt. 11:27; Rom. 11:33-34
- Impartial: Deuteronomy 10:17
- Independent (self-existent): Psalm 115:3, John 5:26; Romans 11:35-36
- Infinite: Psalm 33:11, 90:1-2, 93:2; 145:13; Hebrews 1:8-12
- Jealous: Exodus 34:14; Deuteronomy 4:24; Nahum 1:2; Zechariah 8:2; 2 Corinthians 11:2
- Just: Deuteronomy 32:4; Job 37:23; Psalm 7:11, 99:4; Luke 18:7-8
- Kind: Isaiah 54:8; Jeremiah 9:24; Romans 11:22
- Loving: Exodus 15:13; Psalm 13:5-6, 89:2; Zephaniah 3:17; Romans 8:38-39; Ephesians 3:17-19
- Majestic: Exodus 15:11; Job 37:22; Psalm 8:1, 9; Jude 25
- Merciful (compassionate): 2 Samuel 24:14; Nehemiah 9:31; Psalm 103:11, 145:8-9; Daniel 9:9; Luke 1:50-54
- Mighty: Deuteronomy 10:17; Joshua 4:23-24; Nehemiah 9:32a; Job 9:4; Psalm 18:13-15; 24:8; 93:4; 106:8; 147:5; Isaiah 9:6; Jeremiah 20:11; Luke 1:49; 1 Peter 5:6
- Omnipotent (all-powerful): Genesis 18:14, 1 Samuel 2:6-7; Revelation 19:6
- Omnipresent (present everywhere): Jeremiah 23:23-24; 2 Chronicles: 2:6; Psalm 139:7-16; Acts 17:27-28
- Omniscient (all-knowing): 1 Kings 8:39; Psalm 147:5; Psalm 139:1-6; Proverbs 3:19-20; Jeremiah 32:19; 1 Corinthians 2:10
- One and Only God: Deuteronomy 6:4; Isaiah 45:21-22; 1 Corinthians 8:6; 1 Timothy 2:5
- Patient (longsuffering): Nehemiah 9:30; Romans 3:25; 1 Timothy 1:16; 2 Peter 3:15
- Provider: Philippians 4:19

- Redeemer: Psalm 19:14, 78:35
- Refuge: Psalm 46:1, 91:2
- Righteous: Psalm 11:7; 89:14; Isaiah 51:6;Jeremiah 23:5-6; 1 Corinthians 1:30
- Sacrificial Lamb: John 15:13
- Sovereign: Isaiah 46:10; Psalm 135:6; Daniel 4:35; Ephesians 1:11
- Supreme (pre-eminent): Colossians 1:15-19; Exodus 15: 11, 18; Revelation 19:11-16
- Sustainer: Psalm 3:5; Colossians 1:17
- Transcendent (above and beyond man): Exodus 33:20-23; Job 37:23; Psalm 104:1-4; Ecclesiastes 3:11; Isaiah 40:21-26; 1 Timothy 6:15-16
- Truthful: Numbers 23:19; Psalm 33:4; Isaiah 45:19; John 3:33, 14:6
- Unchanging: Psalm 102:27; Malachi 3:6; James 1:17; Hebrews 13:8
- Unique (incomparable): 1 Kings 8:60; Psalm 89:6; Isaiah 46:9
- Wise: Isaiah 28:29; Jeremiah 10:12; 1 Corinthians 1:30; Colossians 2:2-3[1]

In the next section, I will provide some examples as to how we can see these characteristics of God in His Creation.

Personal Reflection & Application

1. What has the Lord revealed to you by reading this section?

2. Choose three of God's characteristics given in the list provided and consider how you can see them displayed in Creation.

Character of God in Creation

Now that you have a list of God's attributes as a reference point, let me give you some examples that I like to use in my science classes on how we can see those characteristics in Creation. In my classes I teach marine science exclusively so all of these examples are marine-related. However, many of these same characteristics can be applied to other plants and animals. Let the Lord guide you in discovering more creative examples to use in your own science classes.

Example #1 Marine Ecosystems such as Coral Reefs, Seagrass Beds, Rocky Shores, Mangrove Forests

One of the functions of an ecosystem is to provide a residence for different organisms. We can see that regardless of the type of ecosystem, they all provide shelter and a habitat for a variety of animals. We can see the character of God in this because the Bible tells us that God is our shelter, our defense and our protector.

You are my hiding place; You shall preserve me from trouble; You shall surround me with songs of deliverance.

– Psalm 32:7

Just as the Lord is our shelter for those who abide in Him, marine habitats are shelter for animals that live there.

Example #2   A Sacrificial System

There are numerous situations where we see a system of sacrifice in Creation, in which one organism dies so that others may live. This is one of my favorite examples to discuss in class because it is a great opportunity to show my students how all of Creation directly points to Jesus Christ.

Greater love has no one than this, than to lay down one's life for his friends.

– John 15:13

Just one example of a sacrificial system seen in nature, is that seen in the life cycle of a sea palm (*Postelsia palmaeformis*). The sea palm is a type of seaweed that grows attached to rocks along the western coast of North America. Its name is born from its appearance as it looks like a miniature palm tree that only grows to about 2 feet tall. Rather than having a root system like vascular plants, the sea palm is equipped with what is called a "holdfast". This structure at the base of the seaweed allows the sea palm to "grip" the rocks in high energy wave environments. However, the sea palm is not alone in this preferred environment. There are a wide variety of other species that live attached to rocks along the coastline including barnacles, mussels and limpets. Once attached, these species live their entire lives in the same place so competition for an attachment site is fierce.

In order to reproduce, sea palms produce spores that drip down their leaf-like blades onto the substrate at the base of the seaweed. The sea palm only produces spores when it is out of the water so that the spores are not washed away by the tide. If they are able to find an attachment site, these spores eventually mature through the reproductive process into mature adult seaweed. Now, there are some instances in which a sea palm will grow so that its holdfast is straddling a mussel. In this instance the sea palm is not gripping directly onto the rock but onto another organism. When a strong storm increases wave strength, there are times when these mussels and the sea palm attached to them are ripped from the rock by the waves, providing a bare attachment site. The spores left behind by the now-absent adult sea palm and neighboring sea palms are able to monopolize this newly created attachment site and have a greater opportunity for survival. Through the sacrifice of one sea palm, others have the opportunity to survive.[2] In this system of sacrifice, we can see a clear picture of Jesus Christ, the One who gave His life for many.

Example #3 Portuguese man-of-war

Another example as to how we can see the character of God in Creation is through the example of the Portuguese man-of-war and its similarity to the body of Christ.

For as the body is one and has many members, but all the members of that one body, being many, are one body, so also is Christ.

– 1 Corinthians 12:12

When you first see the Portuguese man-of-war, you may think that it is a jellyfish. It looks like a jellyfish as it has a floating soft body and long tentacles but it is not. The Portuguese man-of-war is actually a colony of tiny animals known as hydrozoans. There are four different types of animals that perform different functions in the colony. There are the pneumatophores that make up the colony's float, there are the dactylozooids that make up the tentacles, the gastrozooids that function in feeding and digestion and the gonozooids that function in reproduction. All of the different types of animals are attached to one central column through with they exchange nutrients and information.[3] None of the animals are able to exist without the others as they would be lacking critical functions needed for survival. In this we see a picture of the body of Christ which is described in the Bible being of one body comprised of many members working at different functions. Like the Portuguese man-of-war, all of the members of the body of Christ are important and necessary for survival.

Example #4 Symbiotic relationships

The final example I am going to mention addresses the aspect of friendship. In the Gospel of John, Jesus Christ initiates a dramatic transition in thought as He tells the disciples that they are no longer His servants but His friends.

"No longer do I call you servants, for a servant does not know what his master is doing; but I have called you friends, for all things that I heard from My Father I have made known to you."

– John 15:15

Can you imagine how surprised the disciples must have been to hear the Son of God calling them His friends? What an unlikely friendship! Through Jesus Christ, we can become a friend of the Creator of the Universe! This truth sets the stage for similar unlikely "friendships" seen in all of Creation. These types of "friendships" are known as examples of mutualism, a symbiotic relationship where two organisms work together in order to survive in a mutually beneficial interaction. There are numerous examples that I could mention but I am going to focus on the interaction between the goby (fish) and the shrimp. Both of these animals live in the sandy sea floor near reef environments. Each has physical strengths and weaknesses that are accounted for in their interactions. Their relationship maximizes each animal's strengths while compensating for its weaknesses. The shrimp is an excellent digger and makes swift work of building a burrow to live in. However, his poor eyesight makes him vulnerable to predators. The goby has great eyesight and a good field of vision as his eyes are perched on the top of his head but his fins are poorly used for digging. Consequently, in their interaction with each other, the shrimp digs and maintains a burrow for both animals to live in while the goby keeps a lookout for predators. The shrimp keeps in contact with the goby by touching one of its antennae to the goby's tail. If a threat is sighted, the goby flicks his tail as a warning and both shrimp and goby dive into the safety of their home. This amazing relationship between two unlikely "friends" clearly displays the character of God.

There are numerous more examples that I could share with you but instead I want to encourage you to look at Creation with fresh eyes as you seek to see the character of God in all that He has made. Ask the Lord to reveal to you how you and your children can learn more about Him by what He made. He will surely honor your request!

Personal Reflection & Application

1. What has the Lord revealed to you by reading this section?

2. Can you think of any other examples where you can see one of the characters of God in Creation?

3. How can you teach your children to see the character of God in Creation by adding these exercises to your science class?

Using the Bible in Your Science Classes

In our current culture, we are conditioned to think of religious and academic information as mutually exclusive. However, the Bible transcends this paradigm and is an excellent resource for you to utilize in your science classes for two main reasons.

Reason #1: The Bible must be used as a weapon of spiritual warfare.

Evolutionary science is one of Satan's greatest weapons in drawing people away from the Lord. Using the Bible in your science lessons is an excellent way to combat these attacks against your children. The Bible is clear in describing the Christian walk as a battle against Satan. In John 10:10, Jesus Christ declares Satan's purpose is to "steal, kill and destroy." But yet somehow, we often fail to realize that Satan's intent applies to both adults and children. Perhaps in our minds, we think that Satan will fight "fair" and only target those who are mature enough to fight back. Women of God, Satan has no honor and his desire is to destroy your children. Your role in protecting your children against Satan's attacks while also preparing them to engage in this spiritual battle is crucial.

In the Bible, we frequently see the Apostle Paul describing the Christian walk as engaging in a spiritual battle with Satan. In his letter to the church at Ephesus, Paul makes an analogy comparing aspects of the faith with armor which he calls the "Armor of God." A review of the six pieces of equipment listed will reveal that all of them are defensive in nature with one exception, the "sword of the Spirit, which is the Word of God" (Ephesians 6:17). This is the only offensive weapon discussed. It is critical to fully engage this weapon in your science classes so that you can prepare your children to be victorious in the spiritual battles they will face in their own lives.

Reason #2: The Bible is full of scientific information.

Scientific information in the Bible extends well beyond the account of Creation and it is an excellent supplement to your science lessons. There are numerous instances in which the Bible specifically describes biological processes that we now know through science to be true. Here are just a few for your reference:

- Waves are generated by wind (James 1:6).
- Our earth is surrounded by layers of atmosphere (Amos 9:6).
- The water cycle described in detail (Isaiah 55:10).
- Stars produce sound (Psalm 19:1).

It is important to note the accurate course of events. First, the Bible said it and then scientific studies confirmed its truth. As followers of Jesus Christ, the Bible must be the absolute standard of truth because God is Truth. In instances where science and the Bible appear to contradict, we must align our beliefs with the truth of God's Word.

I once reviewed a textbook used in a marine science class at a local public high school. In one of the chapters, this quote caught my attention: "Science constantly modifies and discards theories and hypotheses based on new information and findings. By constantly proposing, questioning and testing theories and hypotheses, science progresses."[4] Basically, the editors of this secular text book admitted that scientific knowledge in our world is constantly changing. In direct contrast to this statement, we see that the Word of God never changes because our God is the same yesterday, today and forever (Malachai 3:6a and Hebrews 13:8). While the scientific process is honoring to the Lord as it helps for us to gain a greater knowledge of His Creation, the Bible must remain our standard for truth. If there is any instance in which current scientific thought and the Bible contradict, the authority for the follower of Jesus Christ must be the firm foundation of God's Word. Jesus Christ emphasizes this point in His final parable in the Sermon on the Mount.

"Therefore whoever hears these sayings of Mine, and does them, I will liken him to a wise man who built his house on the rock: and the rain descended, the floods came, and the winds blew and beat on that house; and it did not fall, for it was founded on the rock. But everyone who hears these sayings of Mine, and does not do them, will be like a foolish man who built his house on the sand: and the rain descended, the floods came, and the winds blew and beat on that house; and it fell. And great was its fall."

– Matthew 7:24-28

In accordance with this passage, the wise man bases his beliefs on the Rock of biblical truth while the foolish man bases his beliefs on the shifting sand of current scientific thinking.

There are numerous ways that you can include a section on bible study in your children's science classes. For example, you can have them search the Bible for information on Creation and what the Lord says about it. Challenge them to see what scientific discoveries they can find. You can also concentrate your study in a certain area of the Bible. The Psalms are a great place to start as there are numerous chapters praising the Lord for His Creation. You can choose a Psalm a day to review with your children and discuss what you can learn about Creation by reading that Psalm. Here are a few selections to get you started:

- Psalm 8
- Psalm 19
- Psalm 24
- Psalm 29
- Psalm 36
- Psalm 42
- Psalm 65
- Psalm 66
- Psalm 93
- Psalm 104
- Psalm 107
- Psalm 135

In addition, other books like Isaiah or Jeremiah offer wonderful accounts of Creation and God's attributes displayed in Creation. I also encourage my students to memorize passages of scripture that are related to Creation by giving them extra credit for every verse memorized. By incorporating the Bible into your lessons, you are teaching your children that the Word of God applies to all aspects of their lives. As you use the truth of God's Word to supplement your science lessons look for how the Lord works in your children's hearts. The Word of God will not return void and you will enjoy watching your children develop a harvest of righteousness.

Personal Reflection & Application

1. What has the Lord revealed to you by reading this section?

2. What role should the Bible play in your science classes?

3. Why is it significant that the only offensive weapon described in the Armor of God is His Word?

4. Why is it so important to teach children the full counsel of the Word of God?

5. How are you going to incorporate the Bible into your science classes?

Teaching Environmental Stewardship

Earlier in this workbook I discussed the importance of teaching your children to be good stewards of Creation. While being a good steward may look similar to actions of secular environmentalists, it is important to recognize the significant difference in motivation.

So whether you eat or drink or whatever you do, do it all for the glory of God.

– 1 Corinthians 10:31 NIV

As followers of Jesus Christ, we seek to enjoy and utilize natural resources in a manner that brings glory to God. According to the Word of God, we must carefully balance using what we need without exploiting His Creation. Here are some practical suggestions on how you can seek to manage that balance in your household and instill those values in your children.

Try to conserve natural resources by using what you need without being wasteful. Use of critical resources such as fuel, water or energy should be monitored.

- Turn off water or electricity when they are not being used.
- Avoid taking extra long showers.
- Wash dishes in a sink full of water rather than allowing the water to run while washing and rinsing.
- Consider ways that you and your children can reduce errands or driving in the car. Perhaps you can ride bikes, scooters or rollerblade instead.

Consider giving up the convenience of disposable materials in order to help reduce the amount of trash produced by your family. Trash is a significant problem in the marine environment as plastic bags or pieces are eaten by marine life leading to injury, deformities, sickness or death. There is so much trash in the Pacific Ocean that scientists have designated certain areas as garbage patches, particularly in the North Pacific Ocean. These patches are mainly accumulations of small bits of plastic along with other debris that become trapped by ocean currents. These garbage patches vary in size and mass depending on the fluctuations of ocean currents so an exact size estimate is nearly impossible. However, to give you a perspective on size, the area in the North Pacific Ocean that tends to accumulate small pieces of plastic at varying quantities is between 7 to 9 million miles, an area three times the size of the continental United States![5] Clearly, whatever we can do to try and reduce the amount of waste that ultimately may end up in our ocean is important.

While being a good steward may look similar to actions of secular environmentalists, it is important to recognize the significant difference in motivation.

- Try to use reusable materials instead of disposable ones. The greater number of times that you can use something before it is thrown away, the better it is for reducing the amount of waste generated in your area.
- For the products that you do use in your household, try to use as many recyclable products as you can.
    - Disposable water bottles are of particular concern as almost eight out of ten water bottles used in America are thrown away rather than recycled.[6]

    - Avoid materials such as Styrofoam which generally are not accepted at recycle facilities.
- Use reusable canvas or cloth grocery bags instead of plastic or paper bags.
- Minimize the number of disposable cleaning supplies used in your house such as wipes, dust rags, etc.
- Designate reusable water bottles for each member of your family rather than purchasing one-use water bottles and throwing them away.

Discuss with your children how you as a family can be proactive about the growing litter problem in our country.

- Make sure that your family properly disposes of all trash at home and elsewhere in the appropriate waste receptacle.
- Be willing to serve with humble hearts others that are not contentious about proper trash disposal. If you see someone throwing trash on the ground, encourage your children to pick it up for them and throw it away.
- Schedule a waste clean-up event with your homeschool group or friends at a local park or in your community. If you would like to keep it simple, you and your family can do a neighborhood clean-up on your street at regular intervals.

It is probable that none of these ideas are surprising to you as we have all heard similar suggestions from environmentalist groups. Although our actions may look similar to those involved in the green movement, it is imperative that we distinguish ourselves from them by the intent behind what we are doing. Regardless of the actions that you may choose to implement at your house, what is most important is that you encourage your children to make good stewardship a lifestyle out of a desire to honor and serve the Lord.

Personal Reflection & Application

1. What has the Lord revealed to you by reading this section?

2. As followers of Jesus Christ, what should be our motivation behind environmental stewardship?

3. How are you going to implement resource-saving practices in your home as a way of teaching your children to be good stewards of God's Creation?

Beware of Distraction

As this workbook draws to a conclusion, I would like to highlight once again the topic of spiritual warfare. In a previous section, I mentioned that one of your roles in your children's lives is to help protect them from the attacks of Satan. Oftentimes, Satan is most effective influencing our lives by using subtle techniques rather than direct attacks. This is why I would like to bring your attention to the weapon of distraction. Ladies, if you are to effectively teach your children biblically-based curriculum, you must recognize the danger of distraction. If God's purpose for Creation is to encourage your children to understand and know Him, remember this is the last thing that Satan wants. He will try to do everything He can to derail your science lessons with unexpected circumstances and confusion.

> *"The Devil's best trick is to persuade you that he doesn't exist!" – Charles Baudelaire, Le Joueur généreux*

I have seen Satan use the spirit of distraction effectively in many people's lives. So many people today are so distracted by day-to-day life that they have completely forgotten about the spiritual battle being waged around them. Perhaps they have even forgotten about their Enemy. I know that as busy wives, mothers and teachers, you, your home and consequently your children are particularly susceptible to this attack.

How can this manifest itself in your science classes?

- Becoming so consumed with an agenda or deadline that there is no time for your children to engage with the Lord through His Creation.

- Focusing solely on the physical aspects of science in order to meet standardized testing or collegiate requirements.
- Allowing general busyness as a family culture prevent you and your children from taking time to enjoy the blessings the Lord has for you.

Ladies, you must actively guard yourselves and your children against the spirit of distraction. Do not be so task-oriented that you lose the opportunity to utilize your science lessons as a way to facilitate your children's relationship with Jesus Christ. Instead, encourage them to experience Him in a new way through His Creation by creating interactive opportunities for them in the natural environment. While unexpected circumstances will always arise, it is important for you to maintain the proper focus. If you are actively searching for opportunities for your children to engage with the Lord through His Creation, you will find that there is always something available even during your busiest days. Perhaps it is just a few minutes playing in a water fountain in a building courtyard or examining an ant trail on the way to the mailbox. However, even these simple activities can have eternal significance in your children's lives when they are used to point them to their Creator. Be sure that you do not miss them.

Personal Reflection & Application

1. What has the Lord revealed to you by reading this section?

2. How are you going to safeguard against the spirit of distraction in your home?

3. Will you commit to creating a schedule of outdoor activities for your children that will supplement your science class? Use the space provided to draft a schedule that is consistent but also reasonable for your family.

PART 4

In Summary

Summary

1. Dedicate your science classes to Jesus Christ.
    - Display active faith by verbally praying and praising the Lord during your lessons.
    - Remember that science class done correctly is a worship service where praise is directed to the Lord.

2. Maintain the right perspective based on the Word of God.
    - Use the Bible as the foundation for your classes, supplementing curriculum with the Word of God.
    - Teach your children what is true rather than merely focusing on what is not true.
    - Recognize and teach your children to discern the truth behind evolutionary theory. Evolution is a spiritual not a scientific issue.

3. Give your children the opportunity to experience Creation in a hands-on manner.
    - Let them touch, see, feel, hear, taste. Creation is meant to be experienced as a way to interact with our God.
    - Create interactive experiences where they are searching for the character of God.
    - Ask them questions about how they can see the character of God in Creation.
    - Encourage them to think about what they have seen. Give them time to consider and digest what they have experienced.

4. Recognize the danger of distraction.
    - If Creation is meant to encourage you to understand and know God, remember this is the last thing that the Enemy wants.
    - Prepare yourself for spiritual attack.

5. Enjoy the blessing.
    - Our God is a personal God who enjoys displaying His love in Creation for those who seek to honor Him.
    - Remember that the Lord has purposely given us a world that we can understand so that we can come to know Him better. Trust Him to reveal Himself to you and your children through His Creation.
    - As you continue to seek to know the Lord through Creation, He will faithfully continue to reveal Himself to you and your children in new and exciting ways.

Notes

Part 1: Establishing a Biblical Foundation for Your Science Lessons

1. Barna Group (2011) “Six Reasons Young Christians Leave Church” Barna Group. Last updated September 28, 2011. Accessed online February 16, 2013 at http://www.barna.org/teens-next-gen-articles/528-six-reasons-young-christians-leave-church.
2. Pioch, Nicolas (2002) “Giotto di Bondone” WebMuseum, Paris. Last Updated July 27, 2002. Accessed online February 16, 2013 at http://www.ibiblio.org/wm/paint/auth/giotto/.
3. British Broadcasting Corporation News (2012) “Dolphins deserve the same rights as humans, say scientists” BBC News World. Updated February 21, 2012. Accessed online January 28, 2013 at http://www.bbc.co.uk/news/world-17116882.
4. Harrison, Lucy (2010) “How many sharks are killed annually each year?” Shark Specialist Group. Updated August 27, 2010. Accessed online January 28, 2013 at http://www.iucnssg.org/index.php/faqreader/items/how-many-sharks-are-killed-annually-each-year.

Part 2: Addressing the Creation Versus Evolution Issue

1. Carroll, Sean, Ph.D (2005) “Evolution: Constant Change and Common Threads” Lecture Four – From Butterflies to Humans Transcript. Howard Hughes Medical Institute. Updated December 2005. Accessed online January 28, 2013 at http://www.hhmi.org/biointeractive/dvd/transcripts/Evolution%20Lecture%204%20Transcript.pdf
2. DeRosa, Tom (2006) Evolution’s Fatal Fruit. Coral Ridge Ministries, Fort Lauderdale Florida. pg. 169

3. Church of Euthanasia Web Site. Various Articles. Accessed on January 29, 2013 at http://www.churchofeuthanasia.org/index.html.

Part 3: Practical Tips for Your Science Class

1. This list is an expansion of one created by Dr. Kevin Meador. Meador, Dr. Kevin (2005) "Characteristics of God" Stand up for the Truth! Last updated 2005. Accessed online January 31, 2013 at http://standupforthetruth.com/2012/08/to-know-him-is-to-love-him/
2. Attenborough, David (2011) "Palmed Off" The Living Planet. BBC. Digital File. Updated October 12, 2011. Accessed online January 31, 2014 at http://www.bbc.co.uk/programmes/p00l2spt.
3. Bruce, Heather Brinson (2011) "Four in One, One for All: Design in Nature" Answers in Genesis. Updated December 18, 2011. Accessed online February 1, 2012 at http://www.answersingenesis.org/articles/am/v7/n1/four-for-one.
4. Life on an Ocean Planet. (2006) Current Publishing Corp. Margarite, CA. pg. 11-2.
5. NOAA (2012) "De-mystifying the 'Great Pacific Garbage Patch'" National Oceanic and Atmospheric Administration Marine Debris Program. Updated July 19, 2012. Accessed January 31, 2013 at http://marinedebris.noaa.gov/info/patch.html#1.
6. American Chemistry Council (2009) "PET Sales, Recycling and Wasting" Container Recycling Institute. Updated 2009. Accessed January 1, 2013 at http://www.container-recycling.org/index.php/pet-sales-recycling-and-wasting.

About the Author

Christa Jewett is the founder of Saltwater Studies, Inc., a business the Lord led her to begin in 2011. Saltwater Studies offers biblically-based marine science education to homeschooled children and their parents through classes and field trips in South Florida and Phoenix, Arizona. Christa has both a Bachelor of Science and Masters of Science in Marine Biology and 15 years of experience in Marine Sciences. She has worked as a Marine Researcher and an Environmental Consultant studying numerous marine species in locales as varied as Great South Bay, New York, several Caribbean Islands and Queensland, Australia. She has also nearly 10 years of experience in education, specializing in curriculum development, training and instruction of Marine and Environmental Courses. An accomplished writer, she is also a published author and editor of technical articles and educational curriculum. As an instructor, it is Christa's desire to foster in her students a respect and knowledge of the Bible and its scientific accounts, increased knowledge of our Creator and a love for the marine environment. She lives in Fort Lauderdale, Florida with her two dogs, Buster and Gertie.

Contact Information

Christa Jewett, Saltwater Studies
Cell: (954) 895-8579
Email: christa@saltwaterstudies.com
Web: www.saltwaterstudies.com

www.ingramcontent.com/pod-product-compliance
Lightning Source LLC
LaVergne TN
LVHW020644100826
845148LV00012B/2332

* 9 7 8 0 6 1 5 8 1 7 6 0 6 *